ROBOTICS

NANCY DICKMANN

BROWN BEAR BOOKS

Published by Brown Bear Books Ltd
4877 N. Circulo Bujia, Tucson, AZ 85718
USA
and
Studio G14, Regent Studios, 1 Thane Villas,
London N7 7PH, UK

Text: Nancy Dickmann
Design and Illustrations: Square and Circus
Design Manager: Keith Davis
Children's Publisher: Anne O'Daly

Library of Congress Cataloging-in-Publication Data
Names: Dickmann, Nancy, author.
Title: Robotics / by Nancy Dickmann.
Description: Tuscon, AZ : Brown Bear Books Ltd, [2025] | Series: Science starters | Includes bibliographical references and index. | Audience: Ages 6-9 | Audience: Grades 2-3 | Summary: "Find out how robots have developed, how they work, and how people use them"– Provided by publisher.
Identifiers: LCCN 2023052973 (print) | LCCN 2023052974 (ebook) | ISBN 9781781219911 (library binding) | ISBN 9781781219973 (paperback) | ISBN 9781835720035 (ebook)
Subjects: LCSH: Robotics–History–Juvenile literature. | Robots–Juvenile literature.
Classification: LCC TJ211.2 .D53 2025 (print) | LCC TJ211.2 (ebook) | DDC 629.8/92–dc23/eng/20231122
LC record available at https://lccn.loc.gov/2023052973
LC ebook record available at https://lccn.loc.gov/2023052974

Picture Credits
The photographs in this book are used by permission and through the courtesy of:
Alamy: Associated Press 16–17; NASA: JPL-Catech 4–5; Shutterstock: Donatas1205 6–7, Gorodenkoff 18–19, Insta photos 12–13, Miriam Doerr Martin Frommherz 20–21, Pastry Shop 14–15, Maria Sbytova 10–11, Well photo 8–9
All artwork, Brown Bear Books, Shutterstock, and Square and Circus.
t-top, r-right, l-left, c-center, b-bottom

Websites
The website addresses in this book were valid at the time of going to press. However, it is possible that contents or addresses may change following publication of this book. No responsibility for any such changes can be accepted by the author or the publisher. Readers should be supervised when they access the Internet.

Words in **bold** appear in the Glossary on page 23.

Manufactured in the United States of America
CPSIA compliance information: Batch#AG/5660

CONTENTS

WHAT IS A ROBOT?

We use machines every day. They do jobs to make our lives easier. A bicycle is a machine. So are dishwashers and toasters. Each machine does a different job. They are very useful. But they are not robots.

FIGURING IT OUT

Robots take things a step further. They have a computer "brain." They have a set of instructions called a **program**. It tells them what their job is. The robot has to figure out the best way to get it done.

A New Name

"Robota" is a Czech word. It means "forced work." A writer used it in a play in 1920. It was about mechanical workers. That's where our word "robot" comes from.

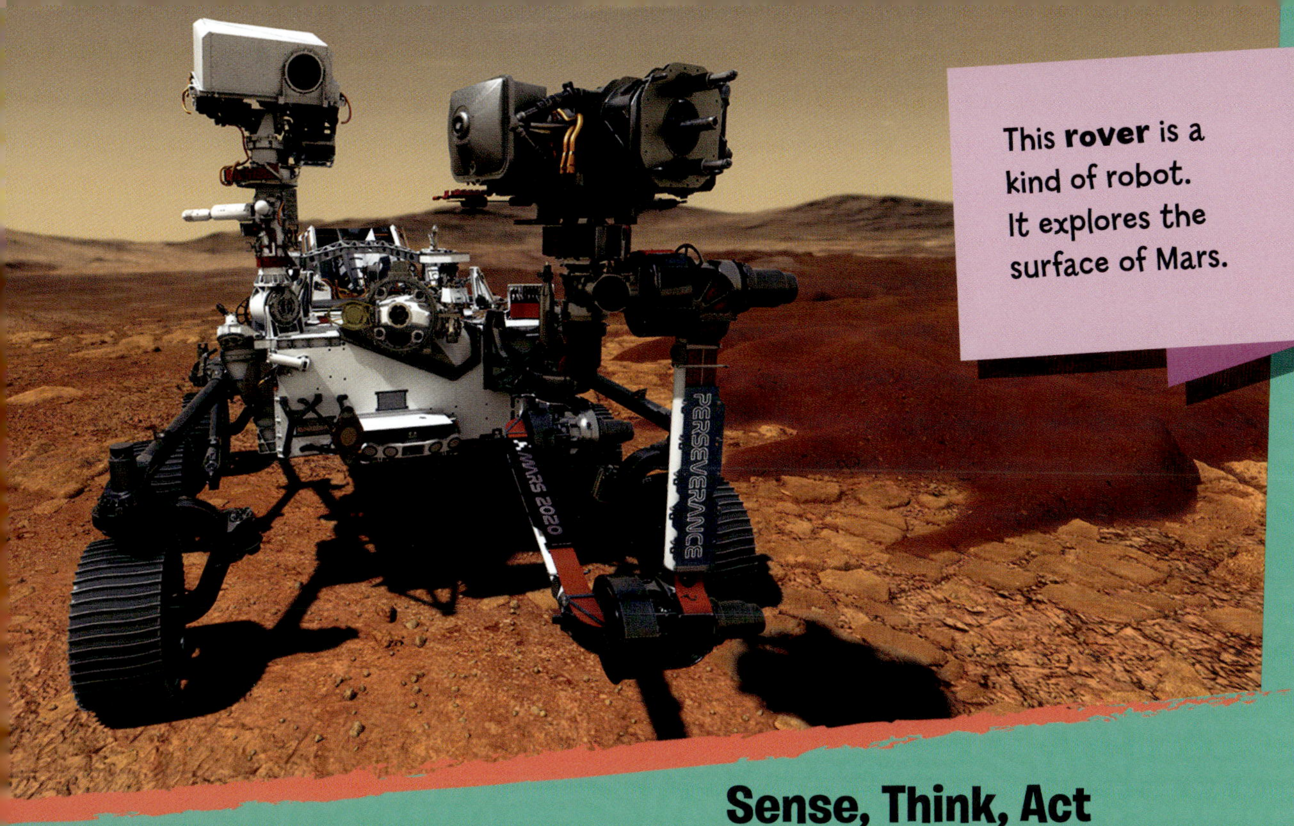

This **rover** is a kind of robot. It explores the surface of Mars.

Sense, Think, Act

Each robot is different. But they all use the same three steps to do a job. Is this how you solve problems?

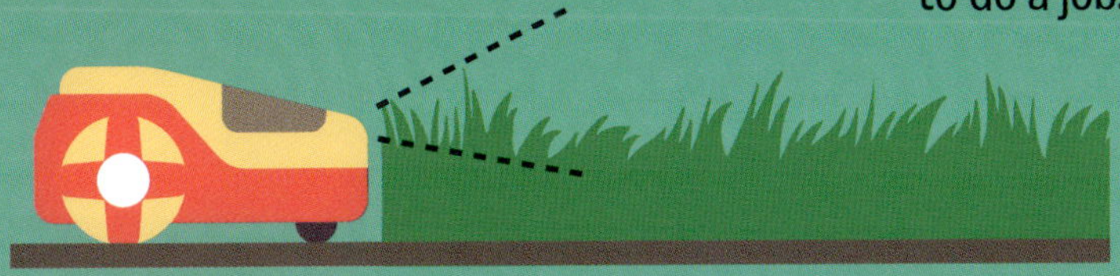

Sense: The robot takes in information. It **senses** the world around it. Is there an obstacle in the way?

Think: The robot thinks about what it has sensed. What's the best way to do the job? Will it need to find a different path?

Act: The robot puts its plan into action. It uses its parts to get the job done.

THE FIRST ROBOTS

Ancient people had machines that moved. The Greeks built moving statues. A Chinese emperor had a mechanical orchestra. In the 1700s people built moving toys. They looked like humans or animals. Some could draw or talk.

REAL ROBOTS

These toys were not true robots. They couldn't be programmed. They were really just wind-up toys. In 1954, the first **factory** robot was invented. It was a mechanical arm. It had a computer program. It could do several different movements.

Old moving toys had metal **gears** inside. They controlled the movement.

Robot Timeline

Robots have come a long way. Their history is short but interesting. Here are some key moments.

1949: Two robot "tortoises" use light to find their way.

1954: The first factory robot is built. It helps build cars.

1969: A better robot arm is designed. It uses electric motors.

1970: A robot rover lands on the moon.

2000: A robot recognizes human emotions. It reacts to them.

2002: The first robot vacuum cleaner is invented.

2005: The first self-driving car is invented.

ROBOTS TODAY

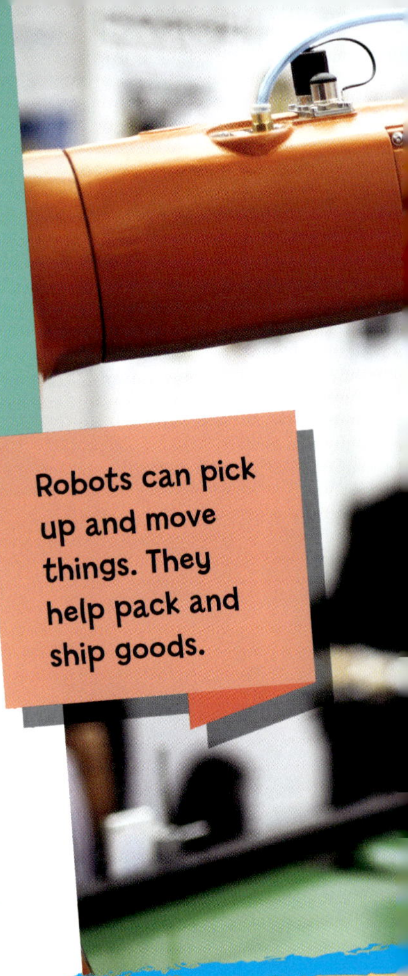

Robots can pick up and move things. They help pack and ship goods.

Today, there are robots everywhere. Factories use robots. So do **warehouses**. There are robots in our homes. Doctors use robots, too. They help with operations. Robots explore the oceans. There are even robots in space!

JOBS FOR ROBOTS

Robots aren't good at football or cooking. But they are great at many other jobs. Robots are tough and strong. A robot can do one thing over and over. It will never get bored or tired. That's why robots are often used in factories.

The four D's

Some jobs are perfect for robots. There are four main types. Robots do things that are dirty or dull. They do jobs that are too dangerous or difficult for us.

Dirty Jobs

- Fixing sewer pipes
- Vacuuming the floor
- Sorting trash

Dull Jobs

- Tightening screws
- Finding things in warehouses
- Restocking shelves in stores

Dangerous Jobs

- Exploring space
- Making bombs safe
- Going into volcanoes

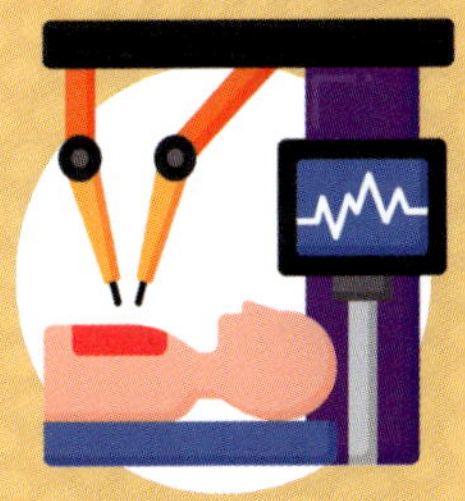

Difficult Jobs

- Making chips for computers
- Doing surgery
- Lifting heavy things

SENSING THE WORLD

When you walk to school, you use your senses. You look for obstacles. You listen for cars. Robots also need to sense the world around them. But they don't have eyes or ears! They use electronic tools instead. These are called **sensors**.

REACTING

What you sense affects how you act. Have you ever touched something hot? You probably pulled your hand away. Robots work the same way. Information goes from a sensor to the computer. It decides how to react.

Powered by Lasers

A movie robot might have **laser** eyes. But some real robots have lasers too! They send out a laser. They wait for it to bounce back. This tells them how far away an object is.

Our senses keep us safe. You see the signal. It tells you that it's safe to cross.

Robot Sensors

Most robots have many sensors. Each one senses a different thing. Here are some of the things a robot's sensors can do.

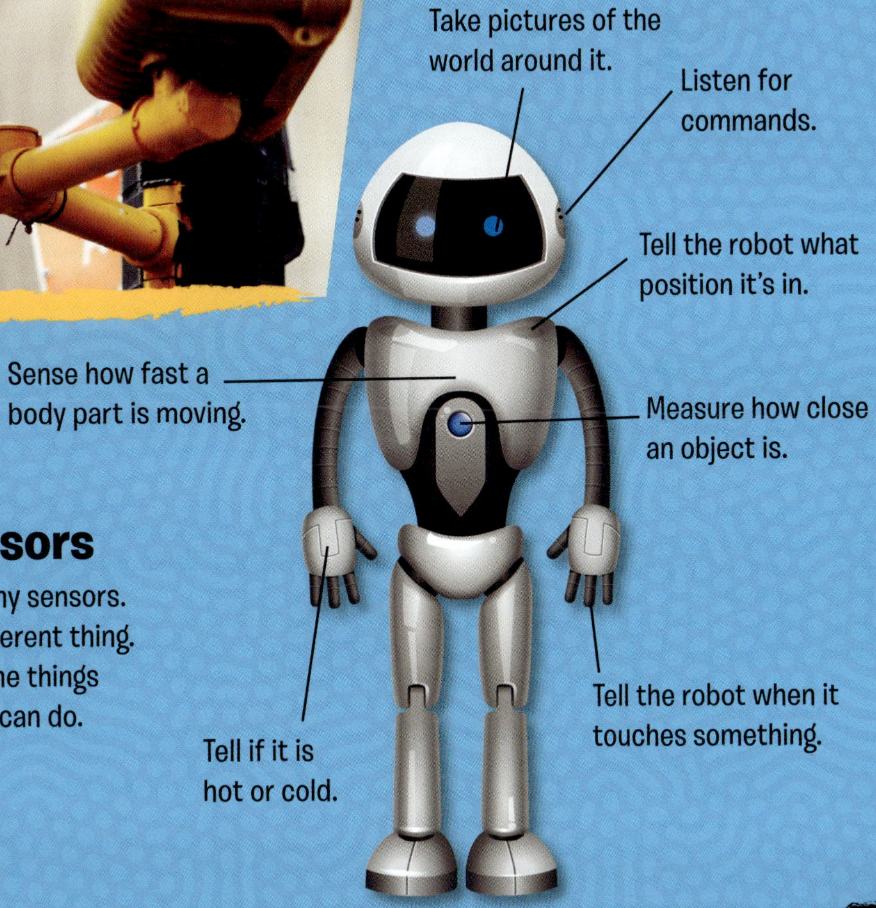

PROGRAMS

A robot has a computer inside. It acts like the robot's brain. The computer gets **data** from the sensors. It controls the robot's body. The computer makes decisions. But it can't do this on its own. It needs instructions.

Writing programs is called coding. Simple coding is easy to learn.

FOLLOWING INSTRUCTIONS

These instructions are the robot's program. A program tells the robot what its job is. It tells it how to solve problems. Programs are written in **code**. Code is made of words, numbers, and symbols. The robot decides what to do. It plans how to do it.

Maze Puzzle

This robot needs to escape the maze. A computer program can tell it what to do.

All you need are four simple commands:

- **fd 1** (move forward one step)
- **bk 1** (move back one step)
- **rt 90** (turn right 90 degrees)
- **lt 90** (turn left 90 degrees)

Could you write a program using these commands? The steps must be in the right order. If not, the robot will get stuck!

ROBOT BODIES

Robots come in all shapes and sizes. Each one is designed for a specific job. It only has the parts that it needs. Some robots can move around. Others are fixed in one place.

MOVING PARTS

Some parts of a robot can move. They have special parts to help them do this. Motors provide the force. **Pistons** move in and out. Parts called servos handle rotation. The computer helps these parts work together. They can raise an arm or move a wheel.

Animal Robots

Animals can move in amazing ways. Some robots copy these skills. There are robots that slither like snakes. Some fly like birds and dragonflies. There is even a hopping kangaroo!

Many factory robots stay in one place. But they can twist and reach.

Robot Shapes

A robot designer chooses the parts that will get the job done.

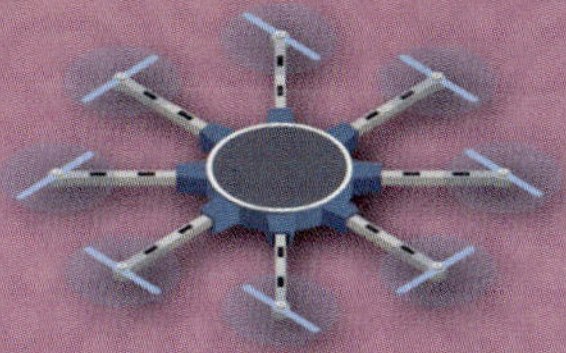

Some robots can fly.

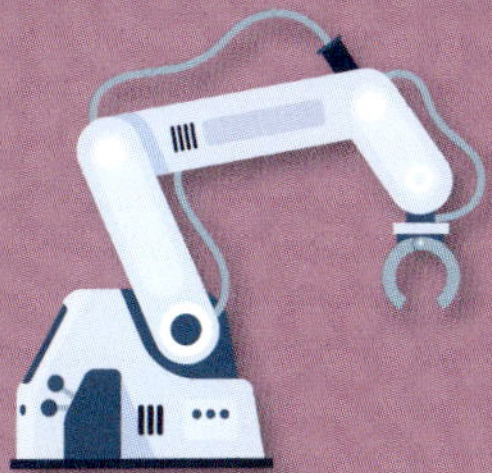

An arm with pincers can pick things up.

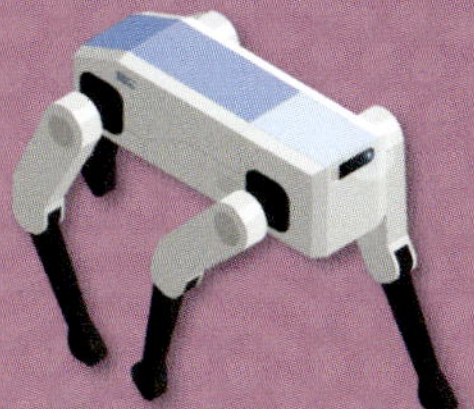

Legs are better for moving on rough ground.

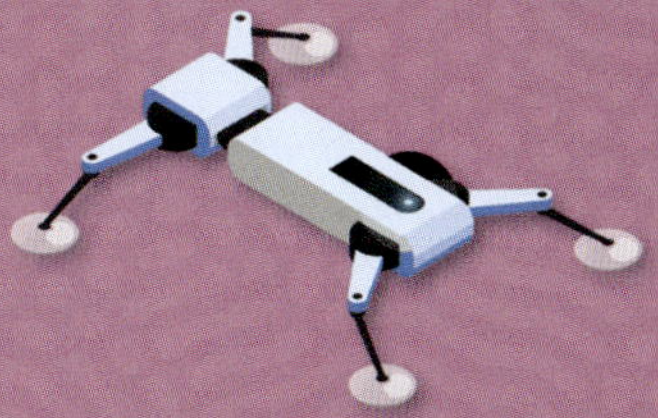

Magnetic feet let a robot climb bridges and buildings.

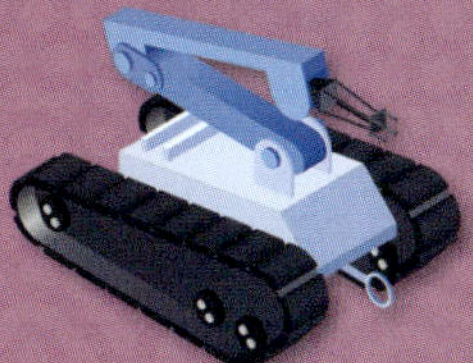

Tracks let a robot roll around.

ANDROIDS

Most robots don't look like humans. But some do. They are called **androids**. Some have a whole body. Others are just a head and shoulders. Some have plastic skin that looks real. Others look more like machines.

JUST LIKE US?

Androids can look very real. But they are not people. Androids can only do what their program tells them to. They cannot love or feel sad. They don't understand right and wrong. Androids can't heal or produce babies. It is hard to get them to work as a team.

The Atlas robot can stand on one leg. It can run, jump, and do handstands!

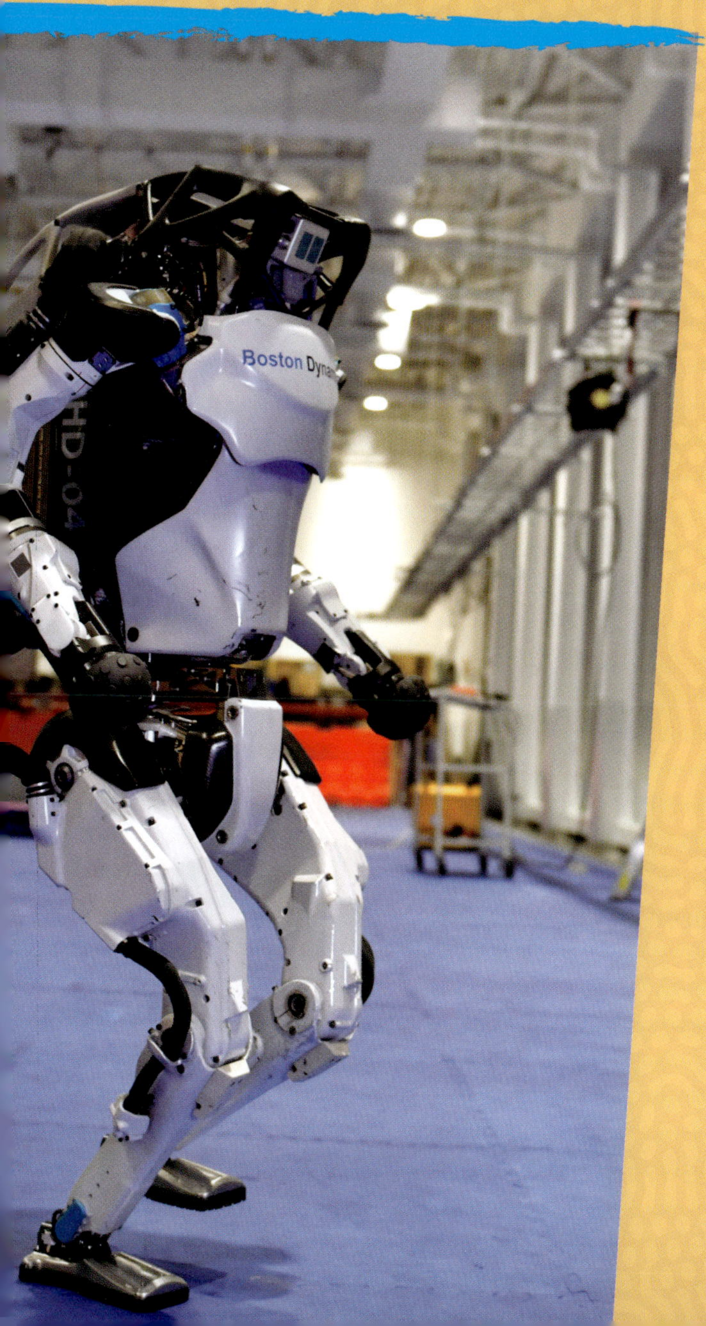

Move Like a Human

Think about all the movements you do in a day. You do most of them without thinking. Some of these are easy for a robot. Others are much harder.

Walking up and down stairs.

Unscrewing a lid.

Writing with a pencil.

Brushing your teeth.

Kicking a ball.

Peeling a banana.

DESIGNING ROBOTS

Every robot is different. They must be carefully designed. A robot designer starts by thinking about the robot's job. They decide what parts it will need. They choose the right sensors. They write a program for it.

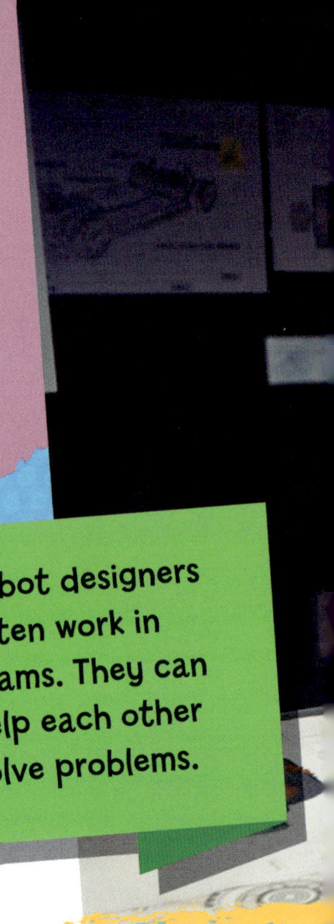

Robot designers often work in teams. They can help each other solve problems.

TEST IT OUT

Designers build a test version of the robot. It's called a **prototype**. It might not be full size. It might not look very polished. But it's a great way to test the design. They see what works and what doesn't. They find solutions and try again.

Robot Contests

Robots can compete against each other! Teams each build a robot for a certain task. It might be playing soccer or climbing stairs. Then they test them. The best robot wins.

From Start to Finish

Inventing a robot is a tough job.
Here are some of the steps.

1. Decide what job the robot will do.
2. How will it move? What parts will it need?
3. What information will it need from its sensors?
4. Sketch out a body. Make sure there's a place for all the parts.
5. Write a program.
6. Build a prototype.
7. Test the prototype.
8. Did it work? Fix any problems and test it again.

ROBOTS OF THE FUTURE

We already use robots. In the future, we will rely on them even more. There might be robots in every home. They could help cook and clean. Robot cars would drive us places. Robot doctors could check our health. Robots could deliver food.

Robots may be able to help elderly people.

OUTSIDE THE HOME

Robots will also work in other places. We will keep using robots to explore space. There will be more robots in factories. Some will make goods. Others will pack and deliver them. Hotels might have robot workers. They will greet guests and clean rooms.

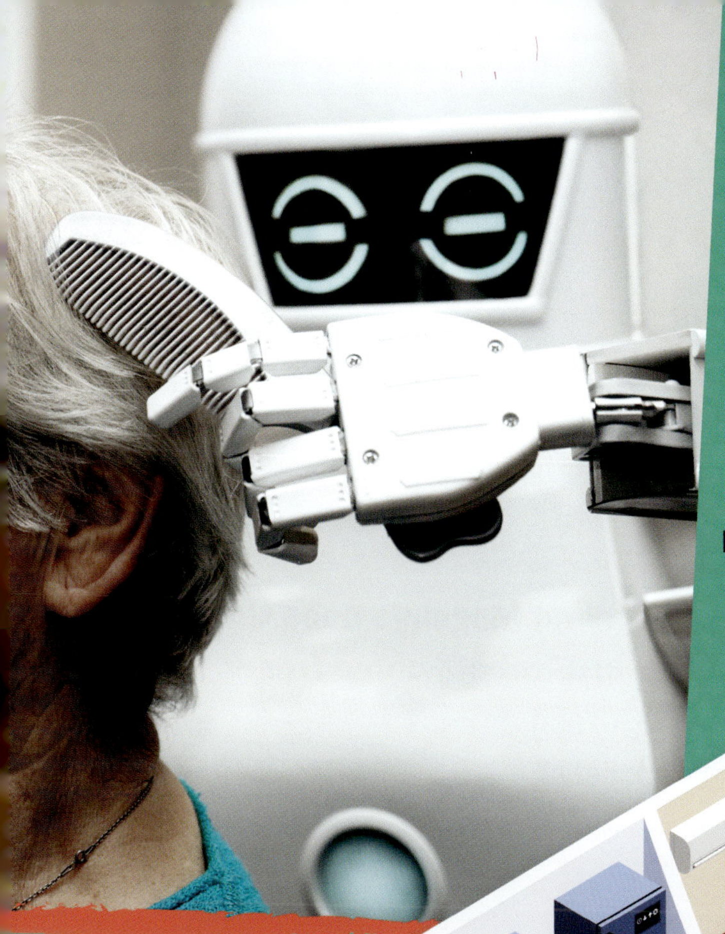

Smart Homes

Robots could make our life easier. They could do many household chores. What will homes look like in the future?

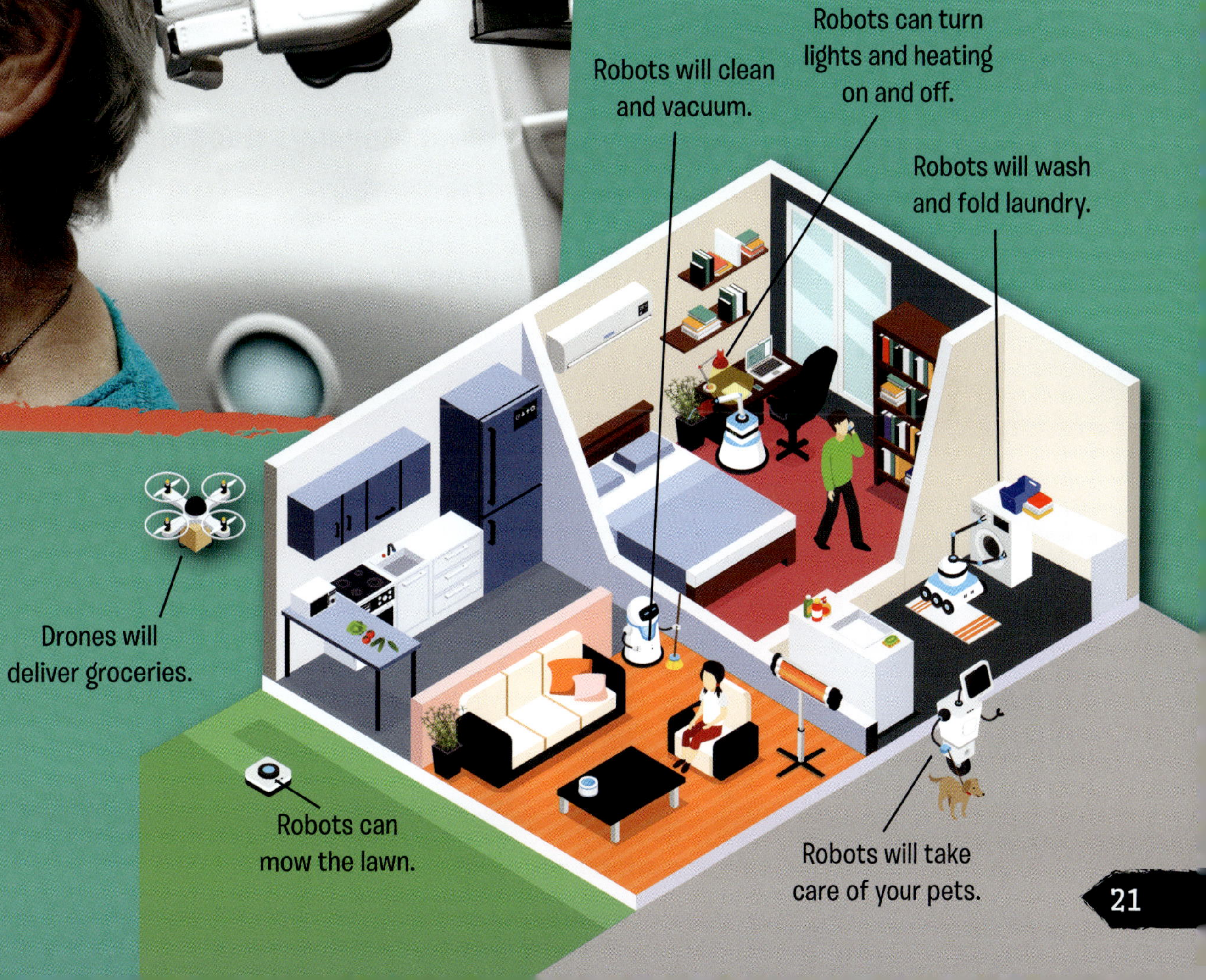

QUIZ

How much have you learned about robotics? It's time to test your knowledge!

1. What do we call a robot's set of instructions?

a. a schedule

b. a program

c. a to-do list

2. What language does the word "robot" come from?

a. Czech

b. Polish

c. Russian

3. What kind of devices help a robot take in information about the world around it?

a. servos

b. sensors

c. pistons

4. What name is given to a simple test version of a new robot?

a. promoter

b. protractor

c. prototype

The answers are on page 24.

GLOSSARY

android a robot that looks like a human

code a system of symbols used as instructions for a computer

data information that is stored or used in a computer, in the form of a series of ones and zeroes

factory a building where products are manufactured

gears metal wheels with teeth around the edges that interlock with each other, so that when one spins around, the others spin too

laser a beam of focused light that can be used to sense distance

piston a metal cylinder that slides up and down inside a tube to make parts of a machine move

program a set of coded instructions for a computer to follow

prototype a simplified test version of a new design

rover a robotic vehicle that drives across the surface of a distant planet or moon

senses physical abilities that let us see, hear, taste, smell, and touch

sensor a device that detects or measures something such as heat or light

warehouse a building where goods are stored until being picked and sent out for delivery

FIND OUT MORE

Books

Household Robots. Sue L. Hamilton, A&D Xtreme, 2019.

Robots (STEAM & Me). Dinah Williams, Starry Forest Books, 2021.

Robots on the Job. Lola Schaefer, Lerner Publications, 2020.

Websites

science.howstuffworks.com/robot.htm

spaceplace.nasa.gov/mars-rovers/en/

www.sciencekids.co.nz/robots.html

INDEX

Answers: 1. b; 2. a; 3. b; 4. c